PI

Series e

Alberto Naranjo C.
Maria del Pilar Noriega E.
Juan Diego Sierra M.
Juan Rodrigo Sanz

Injection Molding Processing Data

2nd Edition

HANSER

Hanser Publications, Cincinnati   Hanser Publishers, Munich

Distributed in North and South America by:
Hanser Publications
6915 Valley Avenue, Cincinnati, Ohio 45244-3029, USA
Fax: (513) 527-8801
Phone: (513) 527-8977
www.hanserpublications.com

Distributed in all other countries by
Carl Hanser Verlag
Postfach 86 04 20, 81631 Munich, Germany
Fax: +49 (89) 98 48 09
www.hanser-fachbuch.de

Cataloging-in-Publication Data is on file with the Library of Congress

ISBN 978-1-56990-666-8
E-Book-ISBN 978-1-56990-667-5

© Carl Hanser Verlag, Munich 2019
Coverdesign: Stephan Rönigk
Typesetted, printed and bound by Kösel, Krugzell
Printed in Germany

Table of Contents

1 Introduction

Initial processing data given to the engineer or technician before setting up an injection molding machine for a new product can save time and money. To arrive at high-quality products as quickly as possible, the machine settings at the beginning of the injection molding process optimization procedure should be as close as possible to optimal processing conditions. However, one needs to be aware that even with a good educated guess of the processing conditions for a given material, the final conditions are dependent on specific material grades, injection molding machine size, screw wear, part and mold design, and other material-independent variables. For most materials a good starting point is always known and can be found in resin supplier data sheets as well as material data banks such as CAMPUS™. This book compiles important processing data information, such as viscosity, thermal properties, mold temperatures, and suggested heater temperatures for the plasticating unit. Through a set of easy to follow examples, this book shows how the given data can be used to generate important information about a specific material, process, and product.

2 Injection Technology

A modern injection molding machine with its most important elements is shown in Figure 2.1. The components of the injection molding machine are the plasticating unit, clamping unit, control unit, and the mold.

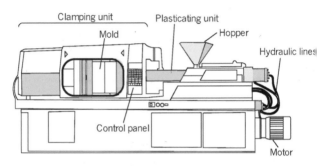

Figure 2.1 Schematic of an injection molding machine. See Figure 2.7 for a more detailed representation of the machine

Today, injection molding machines are classified by the following international convention

Manufacturer type T/P

where T is the clamping force in metric tons and P is defined as

$$P = \frac{v_{max}\ p_{max}}{1000}$$

where v_{max} is the maximum shot size in cm³ and p_{max} is the maximum injection pressure in bar. The clamping force T

an be as low as 1 metric ton for small machines, and as high
s 11,000 tons.

There is another classification regarding specific energy
onsumption (kWh/kg), the Euromap 60.1. There are 10 effi-
iency classes: Class 1 (> 1.5 kWh/kg) to Class 10 (≤ 0.25 kWh/
g). For small machines (screw ≤ 25 mm) the class definition
s different.

2.1 The Injection Molding Cycle

The sequence of events during the injection molding of
plastic part, as shown in Figure 2.2, is called the injec-
ion molding cycle. The cycle begins when the mold closes,
ollowed by the injection of the polymer into the mold
avity. Once the cavity is filled, a holding pressure is main-
ained to compensate for material shrinkage. In the next
tep, the screw turns, feeding the next shot to the front of the
crew. This causes the screw to retract as the next shot is
repared. Once the part is sufficiently cool, the mold opens
nd the part is ejected. Figure 2.3 presents the sequence of
vents during the injection molding cycle. The figure shows
hat the cycle time is dominated by the cooling of the part
nside the mold cavity. However, in some cases the plasti-
ating time can be longer than the cooling time, e.g., when
he mold cavity number is high for the plasticating unit
apacity; the plasticating time is also longer than the cooling
ime when the parts have thin walls. The total cycle time can
e calculated using

$$t_{cycle} = t_{closing} + t_{injectionunitforward} + t_{injection} + t_{cooling} + t_{ejection}$$

where the closing and ejection times, $t_{closing}$ and $t_{ejection}$, can last from a fraction of a second to a few seconds, depending on the size of the mold and the machine.

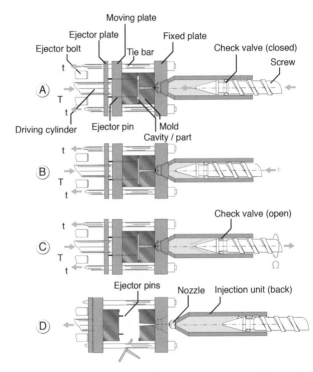

Figure 2.2 Sequence of events during an injection molding cycle

Using the average part temperature history and the cavity pressure history, the process can be followed and assessed using the PVT diagram as depicted in Figure 2.4. To follow the process on the PVT diagram, we must transfer both the temperature and the pressure at matching times. The diagram reveals four basic processes: an isothermal (constant temperature) injection (0–1) with pressure rising to the holding pressure (1–2), an isobaric (constant pressure) cooling process during the holding cycle (2–3), an isochoric (constant volume) cooling after the gate freezes with a pressure drop to atmospheric (3–4), and then isobaric cooling to room temperature (4–5).

The point on the PVT diagram where the final isobaric cooling begins (4) controls the total part shrinkage, Δv. This

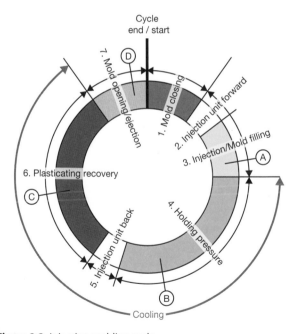

Figure 2.3 Injection molding cycle

point is influenced by the two main processing conditions—the melt temperature, T_M, and the holding pressure, p_H—as depicted in Figure 2.5. Here, the process in Figure 2.4 is compared to one with a higher holding pressure. Of course, there is an infinite combination of conditions that render acceptable parts, bound by minimum and maximum temperatures and pressures.

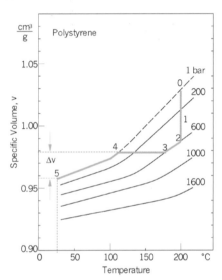

Figure 2.4 Trace of an injection molding cycle in a PVT diagram

Figure 2.6 presents the molding diagram with all limiting conditions. The melt temperature and the injection speed are bound by low values that result in a short shot or unfilled cavity and high values that lead to material degradation. The hold pressure is bound by a low pressure that leads to excessive shrinkage or low part weight, and a high pressure that

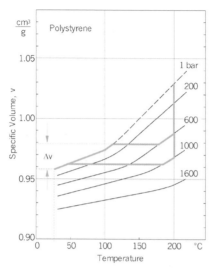

Figure 2.5 Trace of two different injection molding cycles in a PVT diagram

results in flash or jamming. Flash results when the cavity pressure force exceeds the machine clamping force, leading to melt flow across the mold parting line. The holding pressure determines the corresponding clamping force required to size the injection molding machine. An experienced polymer processing engineer can usually determine which injection molding machine is appropriate for a specific application. For the untrained polymer processing engineer, finding this appropriate holding pressure and its corresponding mold clamping force can be difficult. Nowadays there are computer programs for simulation to help them with this critical task. In the following pages some useful equations are presented, and it is explained how to use the physical properties con-

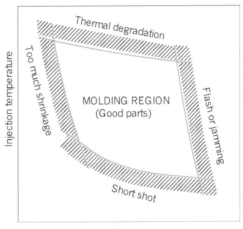

Figure 2.6 The molding diagram

tained in this book to estimate or calculate the most important parameters to set up an injection machine and determine some important values to decide about the right size of the machine.

Determining the setting parameters to operate a mold at one specific machine can be a very difficult task. The same is valid to determine what size or capacity of a machine is required. A unit that is too small for a certain application can delay production of a part and result in increased cost. Knowledge of the required injection pressure, clamping force, and shot size will help when choosing a machine.

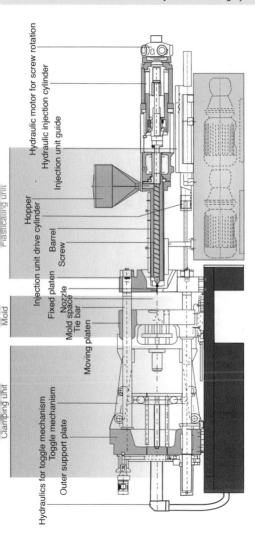

Figure 2.7: Schematic of an injection molding machine

3 Useful Equations and Theory

ESTIMATING COOLING DURING INJECTION MOLDING

The cooling time for a plate-like part of thickness h can be estimated using

$$t_{cooling} = \frac{h^2}{\pi^2 \cdot \alpha} \ln\left(\frac{8}{\pi^2} \frac{T_M - T_W}{T_D - T_W} \right)$$

and for a cylindrical geometry of diameter D using

$$t_{cooling} = \frac{D^2}{23.14\alpha} \ln\left(0.692 \frac{T_M - T_W}{T_D - T_W} \right)$$

In the above equations α represents effective thermal diffusivity, T_M represents the average melt temperature, T_W the average mold temperature, and T_D the average part temperature at ejection.

EQUATIONS FOR PRESSURE FLOW THROUGH A SLIT

Pressure flow through a slit, such as shown in Figure 3.1, is commonly encountered in flows inside injection molds. The Newtonian flow field is described using

$$v_z(y) = \left(\frac{h^2 \Delta p}{8 \mu L}\right)\left[1 - \left(\frac{2y}{h}\right)^2\right]$$

$$Q = \frac{W h^3 \Delta p}{12 \mu L}$$

When using the power law model equation the flow field is described by

$$v_z(y) = \left(\frac{h}{2(s+1)}\right)\left(\frac{h\Delta p}{2mL}\right)^s\left[1 - \left(\frac{2y}{h}\right)^{s+1}\right]$$

$$Q = \frac{Wh^2}{2(s+2)}\left(\frac{h\Delta p}{2mL}\right)^s$$

where $s = 1/n$ and $v_z(y)$ is the velocity profile across the gap and Q the total volumetric flow rate through a slit of width W.

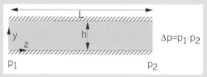

Figure 3.1 Schematic diagram of a pressure flow through a slit

EQUATIONS FOR PRESSURE FLOW THROUGH A TUBE (Hagen-Poiseuille)

Pressure flow through a tube, such as shown in Figure 3.2, is encountered in the runner system in injection molds. The Newtonian flow field is given by

$$v_z(r) = \frac{R^2 \Delta p}{4\mu\, L}\left[1 - \left(\frac{r}{R}\right)^2\right]$$

$$Q = \frac{\pi R^4 \Delta p}{8\mu\, L}$$

Using the power law model, the flow field in a tube is described by

$$v_z(r) = \frac{R}{1+s}\left(\frac{R\Delta p}{2mL}\right)^s\left[1 - \left(\frac{r}{R}\right)^{s+1}\right]$$

$$Q = \left(\frac{\pi R^3}{s+3}\right)\left(\frac{R\Delta p}{2mL}\right)^s$$

The above equations can be used when balancing the runner system inside a multi-cavity injection mold.

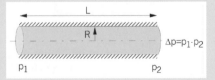

Figure 3.2 Schematic diagram of a tube flow

POWER LAW VISCOSITY MODEL

The power law model is the simplest model that accurately predicts the shear thinning region in the viscosity curve. The model also includes the temperature dependence of the viscosity. The power law model is given by

$$\eta = m_0 e^{-a(T-T_{ref})} |\dot{\gamma}|^{n-1}$$

where m_0 is the consistency index, a is an experimental factor in 1/°C, T is the actual work temperature in °C, $\dot{\gamma}$ is the shear rate in 1/s, and n is the power law index of the fluid.

To evaluate the above equations requires knowledge of the material properties that describe viscosity. Constants that describe the power law viscosity are given in Table 3.1, or can be graphically generated from the viscosity curve as shown later in the examples chapter.

Table 3.1 Power Law and Consistency Indices, Temperature Dependence for Common Thermoplastics

Polymer	m_0 (Pa·s^n)	n	a (1/°C)	T_{ref} (°C)
PS	2.80×10^4	0.28	0.025	170
HDPE	2.00×10^4	0.41	0.002	180
LDPE	6.00×10^3	0.39	0.013	160
PP	7.50×10^3	0.38	0.004	200
PVC	1.70×10^4	0.26	0.019	180

CARREAU VISCOSITY MODEL

The Carreau model is a very common model to describe the viscosity of a polymer melt as a function of shear rate. The model also includes the temperature dependence of the viscosity according to the WLF equation. The model is given by

$$\eta = \frac{A \cdot a_T}{\left(1 + B \cdot \dot{\gamma} \cdot a_T\right)^C}$$

where A is the viscosity at very low shear rate (null viscosity) and B defines the shear rate value where the non-Newtonian behavior begins. T is the actual work temperature in °C, $\dot{\gamma}$ is the shear rate in 1/s, and C is a function ($C = 1-n$) of the power law index n of the fluid.

The factor a_T is defined with the following WLF equation

$$\log a_T = \frac{C_1 \cdot \left(T_0 - T_S\right)}{C_2 + \left(T_0 - T_S\right)} - \frac{C_1 \cdot \left(T_M - T_S\right)}{C_2 + \left(T_M - T_S\right)}$$

where $C_1 = 8.86$, $C_2 = 101.6$, T_S is the standard temperature usually defined as the glass transition temperature $T_G + 30$ K, T_0 the reference temperature, and T_M the actual melt temperature.

ESTIMATING PLASTICATING STROKE

In order to predict the plasticating stroke, L, the following equation can be used

$$L = \frac{4000\,W}{\pi\,D^2\rho}$$

In this equation L is the plasticating stroke (mm), W is the weight (g), D is the diameter of the barrel in the plasticating unit (mm), and ρ the corrected density at the processing temperature (g/cm³). This factor takes into consideration that the screw and barrel are not hermetical at injection time. The parameter ρ can be obtained from Table 3.2.

It is recommended to keep this stroke between D and $3D$. If the stroke is bigger than $3D$, the plasticating unit is too small and the melt quality can be compromised. If the stroke is smaller than D, the plasticating unit is too big, the material can degrade, and the control of the screw position becomes very difficult.

Table 3.2 Density of Several Polymers at Process Temperature

Material	ρ	Material	ρ	Material	ρ
PE	0.71	PS	0.91	PA	0.91
PP	0.73	SB, ABS, SAN	0.88	PC	0.97
PP+20% fillers	0.85	PVC flexible	1.02	PMMA	0.94
PP+40% fillers	0.98	PVC rigid	1.12	POM	1.15
CA	1.02	CAB	0.97	PPO/ PA+fillers	1.06

ESTIMATING SWITCHOVER POSITION

At this position, the screw has injected a volume of melt equal to the volume of the cavity. The volume of the cavity can be easily calculated from the weight to be injected and the specific volume at room temperature taken from the PVT diagram. The following equation can be used

$$L_{sw} = \frac{4 \cdot W \cdot v}{\pi \cdot D^2}$$

In this equation L_{sw} is the travel of screw (mm) from the initial position until the switchover point. W is the weight (g) of melt to be injected into the mold, D is the diameter of the screw in the plasticating unit (mm), and v the specific volume at room temperature (g/cm^3).

ESTIMATING PACKING TIME

During packing time, additional melt is injected into the cavity to compensate for the contraction of the material due to cooling. During this cycle step the temperature starts to decrease. Once the melt temperature is low enough such that the melt can no longer flow (often referred to as no-flow temperature, T_{NF}), the packing pressure has no effect. With the cooling equation, it is possible to calculate the time to reduce the melt temperature to this value (packing time), if instead of T_D the no-flow temperature, T_{NF}, is used. T_{NF} can be taken as the temperature at the crystallization peak from the specific heat capacity diagram, in the case of semi-crystalline materials, or as the glass transition temperature plus 50 K, for amorphous materials.

It is difficult to control and predict the component's shape and residual stresses at room temperature. For example, sink marks in the final product are caused by material shrinkage during cooling, and residual stresses can lead to environmental stress cracking under certain conditions. Warpage in the final product is often caused by processing conditions that lead to asymmetric residual stress distributions through the part thickness. The formation of residual stresses in injection molded parts is attributed to two major coupled factors: cooling and flow stresses. The first and most important is the residual stress formed as a result of rapid cooling.

RESIDUAL STRESSES IN AN INJECTION MOLDED PART

The parabolic temperature distribution through the thickness of a solidified injection molded part leads to a parabolic residual stress distribution, compressive in the outer surfaces of the component and tensile in the core. Assuming no residual stress build-up during the phase change, a simple function based on the parabolic temperature distribution can be used to approximate the residual stress distribution in thin sections:

$$\sigma = \frac{2}{3} \alpha \, E \left(T_s - T_f \right) \left(\frac{6z^2}{4L^2} - \frac{1}{2} \right)$$

Here, T_f is the final temperature of the part, E is the modulus, α the thermal expansion coefficient, L the half thickness, and T_s denotes the solidification temperature: the glass transition temperature for amorphous thermoplastics or the melting temperature for semi-crystalline polymers. Figure 3.3 compares compressive stresses measured and predicted on the surface of PMMA moldings.

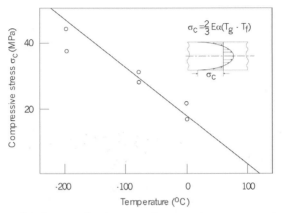

Figure 3.3 Computed and measured compressive stresses on the surface of injection molded PMMA plates

ESTIMATION OF VOLUMETRIC CONTRACTION

The volumetric contraction or shrinkage is estimated by the difference in specific volume between the points 4 and 6, in the PVT diagram, over the specific volume at point 4. The points 4 and 6 are the points where the part stops exerting pressure in the cavity and where the part reaches the room temperature after it is ejected, respectively.

$$S_v = \frac{v_o - v_f}{v_0} 100\%$$

In the above equation, S_v is the volumetric contraction or shrinkage expressed in %, v_0 is the specific volume at point 4 in cm³/g, and v_f is the specific volume at point 6 in cm³/g.

TEMPERATURE PROFILE OF INJECTION MOLDING MACHINES

The temperature profile of an injection molding machine should be modified according to the plasticating stroke, for an appropriate residence time. If the plasticating stroke is under 50% of the maximum stroke (normally 3D), the recommended temperature profile is ascending. If the plasticating stroke is over 50% of the maximum stroke, the recommended temperature profile is descending as can be seen in Figure 3.4.

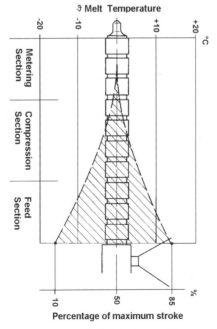

Figure 3.4 Temperature profile in injection molding machines

4 Examples

4.1 Runner System Balance

As an example, let us consider the multi-cavity injection molding process shown in Figure 4.1. To achieve equal part quality, the filling time for all cavities must be balanced. For the case in question, we need to balance the cavities by solving for the runner radius, R_2. For a balanced runner system the flow rates into all cavities must match. For a given flow rate Q, length L, and radius R_1, we can also solve for the pressures at the runner system junctures. Assuming an isothermal flow of a non-Newtonian shear thinning polymer with viscosity η we can compute the radius R_2 for a part molded of polystyrene with a consistency index of 2.8×10^4 Pa·s^n and a power law index (n) of 0.28. The flow through each runner section is governed by the flow-through-a-tube equation, and the various sections can be represented using

Section 1: $\quad 4Q = \left[\dfrac{\pi \, (1.5R_1)^3}{s+3} \right] \left[\dfrac{1.5R_1 \left(p_1 - p_2 \right)}{2mL} \right]^s$

Section 2: $\quad 2Q = \left[\dfrac{\pi \, (1.5R_1)^3}{s+3} \right] \left[\dfrac{1.5R_1 \left(p_2 - p_3 \right)}{2mL} \right]^s$

Section 3: $\quad Q = \left[\dfrac{\pi \, R_2^3}{s+3} \right] \left[\dfrac{R_2 \left(p_2 - 0 \right)}{2m(2L)} \right]^s$

Section 4: $\quad Q = \left[\dfrac{\pi \, R_1^3}{s+3} \right] \left[\dfrac{R_1 \left(p_3 - 0 \right)}{2m(2L)} \right]^s$

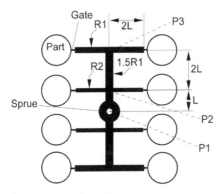

Figure 4.1 Runner system layout

Using values of $L = 10$ cm, $R_1 = 3$ mm, and $Q = 20$ cm³/s, the unknown parameters, P_1, P_2, P_3, and R_2, can be obtained using the above equations. The equations are non-linear and must be solved in an iterative manner. For the given values, a radius, R_2, of 2.61 mm would result in a balanced runner system, with pressures $P_1 = 478.1$ bar, $P_2 = 376$ bar, and $P_3 = 292$ bar. For comparison, if one had assumed a Newtonian model with the same consistency index and a power law index of 1.0, a radius R_2 of 2.76 mm would have resulted, with much higher required pressures of $P_1 = 3,847$ bar, $P_2 = 2,456$ bar, and $P_3 = 1,760.5$ bar. The difference is due to shear thinning.

4.2 Shrinkage Prediction

A thin polyamide 66 component is injection molded under the following conditions:

▶ The melt is injected at 275 °C to a maximum pack/hold pressure of 800 bar

▶ The 800 bar pack/hold pressure is maintained until the gate freezes off, at which point the part is at an average temperature of 175 °C

▶ The pressure drops to 1 bar as the part cools inside the cavity

▶ The part is removed from the mold and cooled to 25 °C

Draw the entire process on the PVT diagram. Estimate the final part thickness if the mold thickness is 1 mm. For thin injection molded parts, most of the shrinkage leads to part thickness reduction.

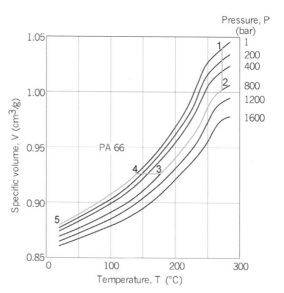

Figure 4.2 PVT diagram for PA66

Both, the specific volume when the mold is opened and the specific volume when the part is cold, are obtained from the cycle traced in the above PVT diagram. The specific volume when the mold opens is $v_0 = 0.93$ cm³/g and the final volume is $v_f = 0.88$ cm³/g. The shrinkage of the injected piece is calculated using

$$S_v = \frac{v_o - v_f}{v_0} 100\% = 5.38\%$$

and the final thickness e_f is calculated by

$$e_f = \frac{v_f}{v_0} e_o = \frac{0.88}{0.93} 1 \text{ mm} = 0.9462 \text{ mm}$$

4.3   Cooling Time Prediction

Estimate the cooling time for an LDPE product with a thickness of 3 mm, for a mold temperature of 30 °C.

Figure 4.3 graphs the cooling time versus thickness using the equation of cooling time for plate-like parts, assuming typical melt and ejection temperatures of LDPE. In this graph, for a 3 mm thickness part and a cavity temperature of 30 °C, the cooling time is 22 s.

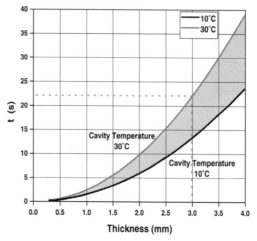

Figure 4.3 Cooling time versus thickness for LDPE plate-like parts

4.4 Effect of Process Parameters on Shrinkage

As a consequence of the increase in injection pressure, the volumetric contraction can be significantly reduced. Two injection molding cycles at different injection pressures, for polystyrene Hostyren Type S3200, are shown in the PVT diagram of Figure 4.4. When the process pressure is increased from 200 bar to 400 bar, the volumetric shrinkage is reduced from 5 to 3%, which means a 40% reduction.

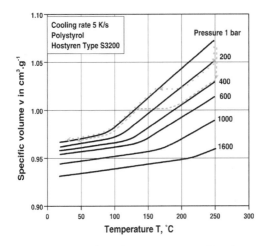

Figure 4.4 PVT diagram for polystyrene Hostyren Type S3200

4.5 Theoretical Energy Consumption in an Injection Molding Process with PS

Estimate the theoretical energy consumption for processing PS from room temperature (25 °C) to 200 °C and the theoretical amount of heat that has to be removed in the mold if the ejection temperature is 40 °C.

The theoretical energy consumption for processing PS from 25 to 200 °C can be calculated by reading the specific enthalpy at 25 and 200 °C, respectively.

$$Energy = \Delta H_{200°C} - \Delta H_{25°C} = 321\frac{kJ}{kg} - 18.2\frac{kJ}{kg} = 302.8\frac{kJ}{kg}$$

The heat removed in the mold from the melt temperature (200 °C) to an ejection temperature of 40 °C can be calculated by reading the specific enthalpy at 40 and 200 °C, respectively

$$Heat = \Delta H_{200°C} - \Delta H_{40°C} = 321\frac{kJ}{kg} - 37.3\frac{kJ}{kg} = 283.7\frac{kJ}{kg}$$

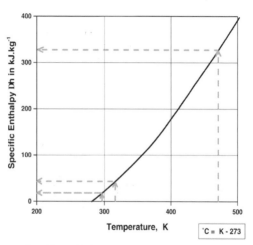

Figure 4.5 Specific enthalpies

4.6 Processing Conditions of PA6 Parts and Theoretical Power-Cooling Requirement

A part of PA6 with 2 mm average thickness and 50 g will be produced by injection molding. What would the processing conditions be for producing the part? Estimate the theoretical power-cooling requirement.

Recommended processing conditions: from the mold cooling diagram (Figure 5.50):

Melt temperature	260 °C
Ejection temperature	90 °C

From the recommended temperature profiles for injection molding of PA6 (Figure 5.51):

Zone	Temperature
1 (bin)	210 to 230 °C
2	215 to 235 °C
3	220 to 240 °C
4	225 to 245 °C
5	230 to 250 °C
6 (die)	230 to 250 °C

Cavity temperature: 70 °C (recommended: 60 to 80 °C). From the appendix, the following pre-drying conditions can be obtained for PA6:

Drying temperature	75 to 85 °C
Drying time	2 to 20 hours
Maximal water content	0.1 to 0.2 wt %

Cooling Time

From the mold cooling diagram, with a cavity temperature of 70°C, a melt temperature of 260°C, an ejection temperature of 90°C, and a thickness of 2 mm:

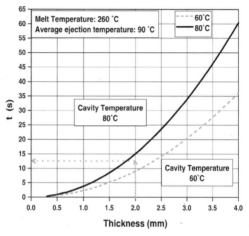

Figure 4.6 Cooling time

Cooling time	12 seconds
Cycle time	12 seconds + 4 seconds (estimated) = 16 seconds

Theoretical Power Required for Mold Cooling

The heat removed in the mold could be calculated from the enthalpy diagram:

$$Heat = \left(\Delta H_{90^\circ C} - \Delta H_{260^\circ C}\right)W = \left(109\frac{kJ}{kg} - 618\frac{kJ}{kg}\right)\frac{50}{1000}\,kg$$

$$Heat = -25.45\ kJ$$

The theoretical power required for mold cooling could be calculated from:

$$Theoretical\ Power = \frac{25.45\ kJ}{16\ s} = 1.59\ kW$$

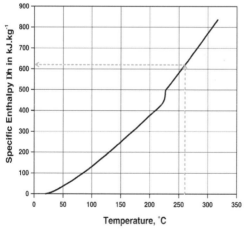

Figure 4.7 Specific enthalpy for PA6

4.7 Theoretical Water Flow Requirement for a Mold for PA6 Parts

The above calculated theoretical power for mold cooling represents the energy to be transported by the cooling medium during the cycle time, if the heat exchange with the surroundings is neglected. The following equation is used to estimate the necessary flow of cooling medium to fulfill this task:

$$\frac{Heat}{t_{cycle}} = \frac{m_{coolant}}{t_{cycle}} \cdot \Delta h_{coolant} = \left(\dot{V}_{coolant} \cdot \rho_{coolant} \right) \cdot \left(c_{p_{coolant}} \cdot \Delta T_{coolant} \right)$$

where $m_{coolant}$ is the mass of water flow and $\Delta h_{coolant}$ the specific enthalpy change during the cycle time t_{cycle}. $\dot{V}_{coolant}$ is the volume flow of cooling medium; $\rho_{coolant}$ and $c_{p_{coolant}}$ its density and specific heat capacity, respectively. $\Delta T_{coolant}$ is the temperature change of cooling medium between inlet and outlet. Reordering the terms we obtain the expression to calculate the cooling flow rate:

$$\dot{V}_{coolant} = \frac{Heat}{t_{cycle} \cdot \rho_{coolant} \cdot c_{p_{coolant}} \cdot \Delta T_{coolant}}$$

Continuing with the example above, for a typical allowable temperature increase of 1.5 K for cooling water, with a water density of 1 kg/l and a specific heat of 4.8 kJ/kg·K, the following value for the flow rate of cooling water is obtained

$$\dot{V}_{coolant} = \frac{25.45 \text{ kJ}}{16 \text{ s} \cdot 1 \dfrac{\text{kg}}{\text{l}} \cdot 4.8 \dfrac{\text{kJ}}{\text{kg} \cdot \text{K}} \cdot 1.5 \text{ K}} = 0.22 \frac{\text{l}}{\text{s}} = 13.3 \frac{\text{l}}{\text{min}}$$

4.8   Determining Viscosity Constants from the Curve

Taking the viscosity for LDPE in the polymer data chapter of this book, we first determine the power law index by setting the slope of the curves equal to $n-1$. Note that the scales must be the same. If the scales are off by a factor, the slope must be modified accordingly. In Figure 4.8 we simply distorted the scales of the curve given in the polymer data chapter. For a shear rate of 1.0 s^{-1} and a temperature equal to the reference temperature of 190 °C, we solve for the consistency index. It is now possible to set the temperature to 160 °C and using the same conditions we can solve for the temperature dependence constant.

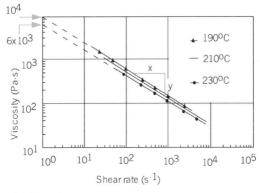

Figure 4.8 Viscosity curve

$$n-1 = \frac{-y}{x}0.61 \quad \Rightarrow \quad n = 0.39$$

$$m_0 = 1\times10^4 \ \text{Pa}\cdot\text{s}\left(T_{\text{ref}} = 190 \ ^\circ\text{C}\right)$$

$T = 230\ ^{\circ}\text{C}$

$\eta = 6 \times 10^3\ \text{Pa} \cdot \text{s}$

$$\frac{6 \times 10^3\,\text{Pa} \cdot \text{s}}{1 \times 10^4\,\text{Pa} \cdot \text{s}} = e^{-a(230\ ^0C - 190\ ^0C)} \Rightarrow a = 0.13/^0\text{C}$$

5 Polymer Data (for Standard Materials without Fillers or Modifiers)

5.1 Polyolefins

5.1.1 Low Density Polyethylene (LDPE)

Basic technical data

▸ Density: 0.910 to 0.926 g/cm³
▸ Melting point: 105 to 115 °C
▸ Glass transition temperature: −133 to −120 °C

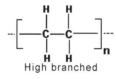

High branched

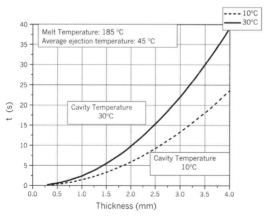

Figure 5.1 Mold cooling with LDPE

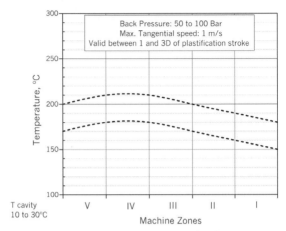

Figure 5.2 Recommended temperature profiles for processing LDPE

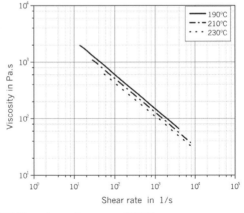

Figure 5.3 Viscosity vs shear rate of LDPE

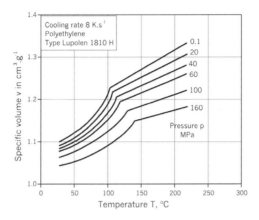

Figure 5.4 PVT diagram for LDPE

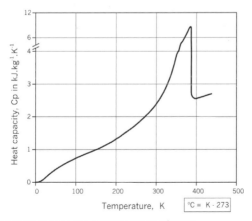

Figure 5.5 Heat capacity vs temperature of LDPE

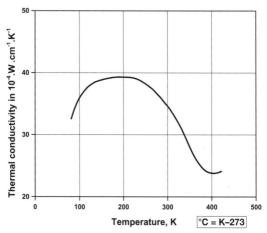

Figure 5.6 Conductivity vs temperature of LDPE

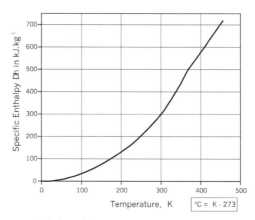

Figure 5.7 Enthalpy of LDPE

5.1.2 High Density Polyethylene (HDPE)

Basic technical data

▸ Density: 0.940 to 0.972 g/cm^3
▸ Melting point: 130 to 135 °C
▸ Glass transition temperature: −120 to −90 °C

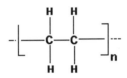

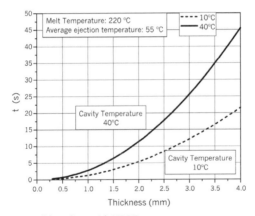

Figure 5.8 Mold cooling with HDPE

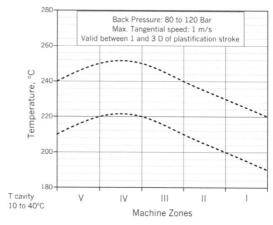

Figure 5.9 Recommended temperature profiles for processing HDPE

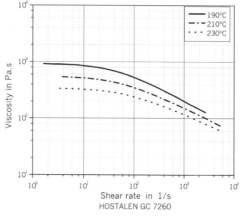

Figure 5.10 Viscosity vs shear rate of HDPE

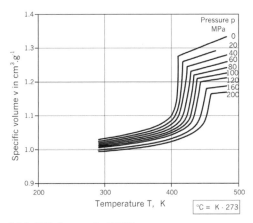

Figure 5.11 PVT diagram for HDPE

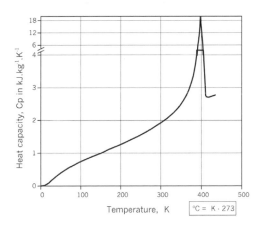

Figure 5.12 Heat capacity vs temperature of HDPE

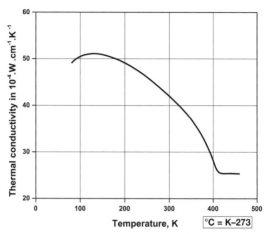

Figure 5.13 Conductivity vs temperature of HDPE

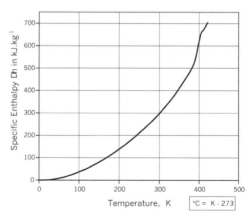

Figure 5.14 Enthalpy of HDPE

5.1.3 Polypropylene Homopolymer (PP)

Basic technical data

- Density: 0.900 to 0.910 g/cm³
- Melting point: 165 to 176 °C
- Glass transition temperature: −10 to 0 °C

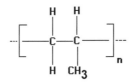

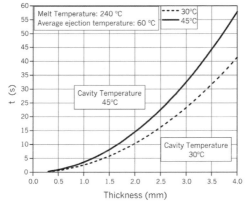

Figure 5.15 Mold cooling with PP

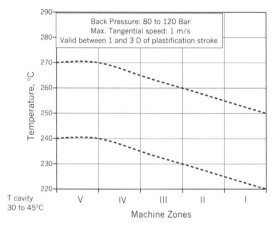

Figure 5.16 Recommended temperature profiles for processing PP

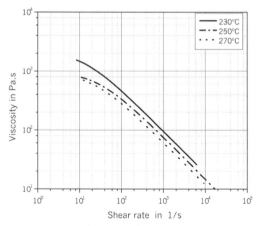

Figure 5.17 Viscosity vs shear rate of PP

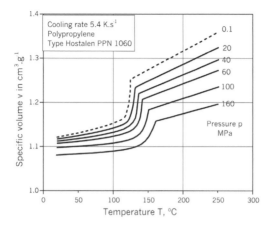

Figure 5.18 PVT diagram for PP

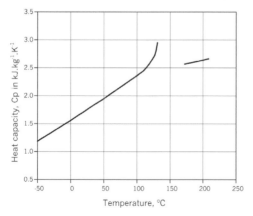

Figure 5.19 Heat capacity vs temperature of PP

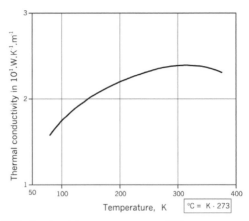

Figure 5.20 Conductivity vs temperature of PP

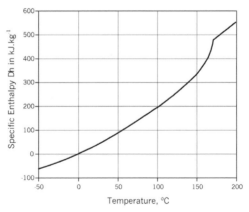

Figure 5.21 Enthalpy of PP

5.2 Styrenics

5.2.1 Polystyrene Homopolymer (PS)

Basic technical data

- Density: 1.040 to 1.065 g/cm^3
- Amorphous
- Glass transition temperature: 80 to 113 °C

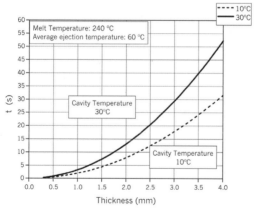

Figure 5.22 Mold cooling with PS

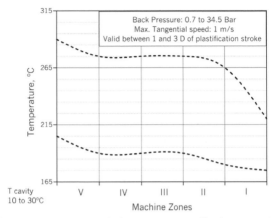

Figure 5.23 Recommended temperature profiles for processing PS

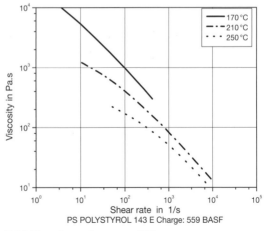

Figure 5.24 Viscosity vs shear rate of PS

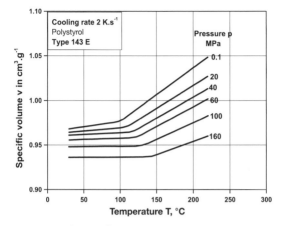

Figure 5.25 PVT diagram for PS

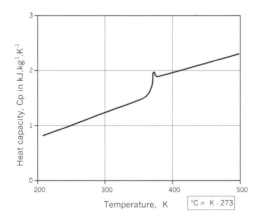

Figure 5.26 Heat capacity vs temperature of PS

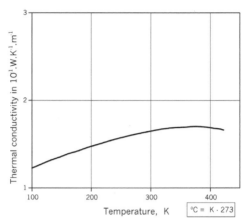

Figure 5.27 Conductivity vs temperature of PS

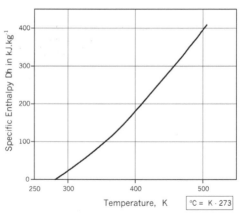

Figure 5.28 Enthalpy of PS

5.2.2 High Impact Polystyrene (HIPS)

Basic technical data

- Density: 1.03 to 1.05 g/cm³
- Amorphous
- Glass transition temperature (continuous phase): −60 to −20 °C

Chemical formula: Heterophasic material

- Matrix: polystyrene
- Disperse phase:

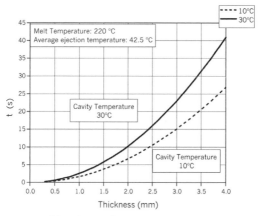

Figure 5.29 Mold cooling with HIPS

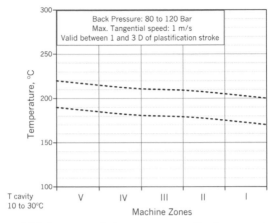

Figure 5.30 Recommended temperature profiles for processing HIPS

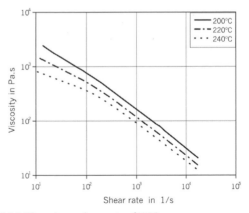

Figure 5.31 Viscosity vs shear rate of HIPS

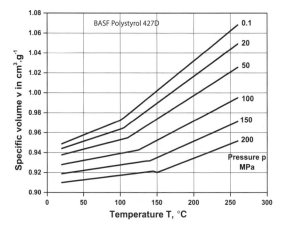

Figure 5.32 PVT diagram for HIPS

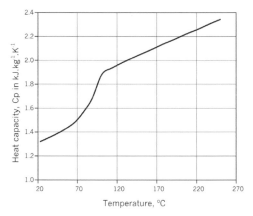

Figure 5.33 Heat capacity vs temperature of HIPS

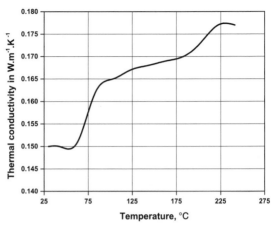

Figure 5.34 Conductivity vs temperature of HIPS

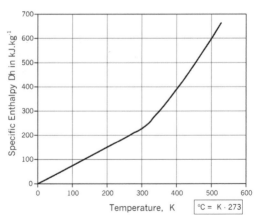

Figure 5.35 Enthalpy of HIPS

5.2.3 Styrene/Acrylonitrile Copolymer (SAN)

Basic technical data

- Density: 1.07 to 1.09 g/cm^3
- Amorphous
- Glass transition temperature: 95 to 105 °C

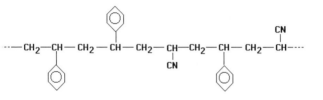

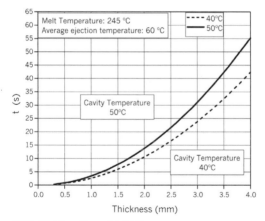

Figure 5.36 Mold cooling with SAN

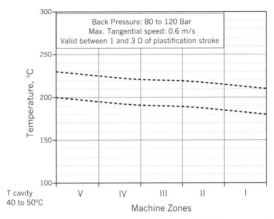

Figure 5.37 Recommended temperature profiles for processing SAN

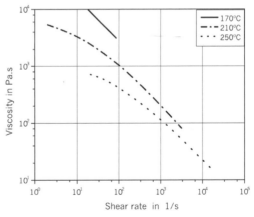

Figure 5.38 Viscosity vs shear rate of SAN

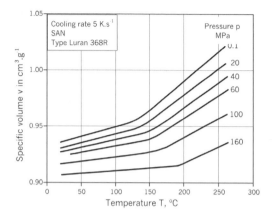

Figure 5.39 PVT diagram for SAN

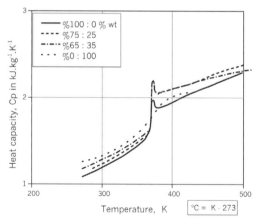

Figure 5.40 Heat capacity vs temperature of SAN

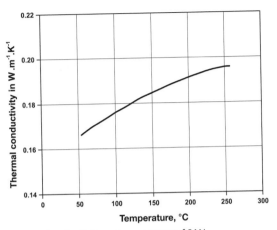

Figure 5.41 Conductivity vs temperature of SAN

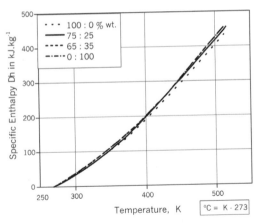

Figure 5.42 Enthalpy of SAN

5.2.4 Acrylonitrile/Butadiene/Styrene Copolymer (ABS)

Basic technical data

- Density: 1.040 to 1.060 g/cm^3
- Amorphous
- Glass transition temperature: 80 to 125 °C

Chemical formula: Heterophasic material

- Matrix: SAN
- Disperse phase: poly(butadiene-g-(styrene-co-acrylonitrile))

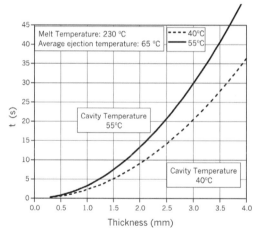

Figure 5.43 Mold cooling with ABS

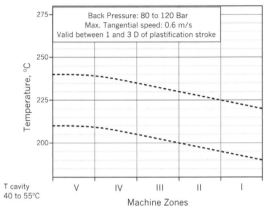

Figure 5.44 Recommended temperature profiles for processing ABS

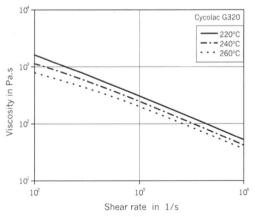

Figure 5.45 Viscosity vs shear rate of ABS

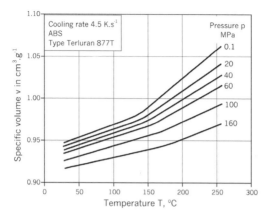

Figure 5.46 PVT diagram for ABS

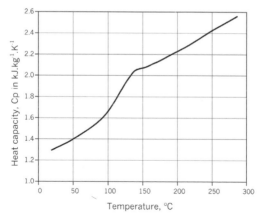

Figure 5.47 Heat capacity vs temperature of ABS

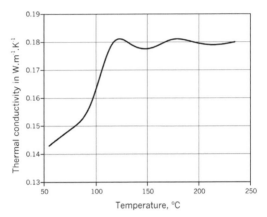

Figure 5.48 Conductivity vs temperature of ABS

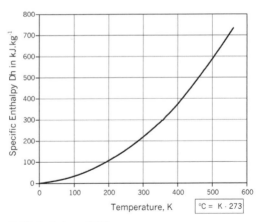

Figure 5.49 Enthalpy of ABS

5.3 Polycondensates

5.3.1 Polyamide 6 (PA6)

Basic technical data

▶ Density: 1.120 to 1.150 g/cm³
▶ Melting point: 180 to 225 °C
▶ Glass transition temperature: 60 °C

$$[-NH(CH_2)_5CO-]\ldots$$

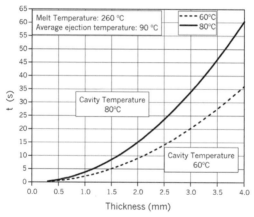

Figure 5.50 Mold cooling with PA6

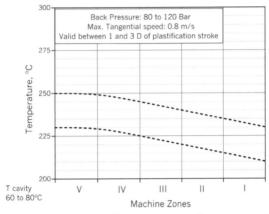

Figure 5.51 Recommended temperature profiles for processing PA6

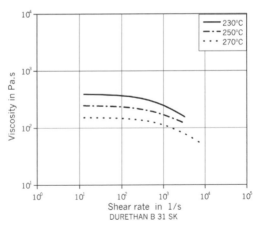

Figure 5.52 Viscosity vs shear rate of PA6

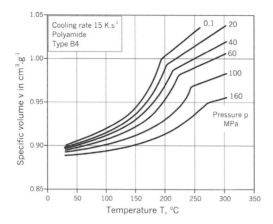

Figure 5.53 PVT diagram for PA6

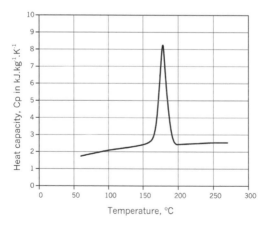

Figure 5.54 Heat capacity vs temperature of PA6

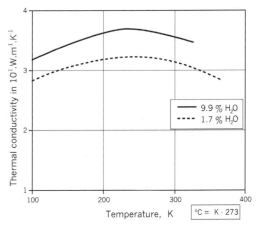

Figure 5.55 Conductivity vs temperature of PA6

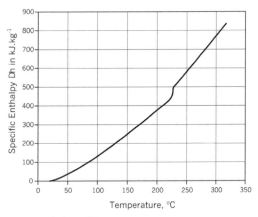

Figure 5.56 Enthalpy of PA6

5.3.2 Polyamide 66 (PA66)

Basic technical data

▶ Density: 1.130 to 1.160 g/cm³
▶ Melting point: 225 to 250 °C
▶ Glass transition temperature: 70 °C

$$[-NH(CH_2)_6NH-CO(CH_2)_4CO-]\dots$$

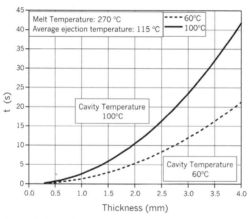

Figure 5.57 Mold cooling with PA66

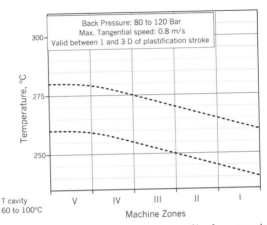

Figure 5.58 Recommended temperature profiles for processing PA66

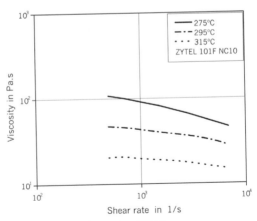

Figure 5.59 Viscosity vs shear rate of PA66

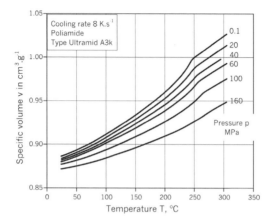

Figure 5.60 PVT diagram for PA66

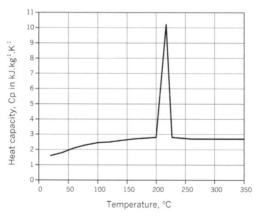

Figure 5.61 Heat capacity vs temperature of PA66

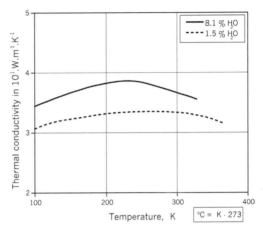

Figure 5.62 Conductivity vs temperature of PA66

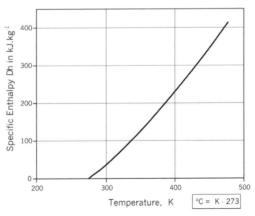

Figure 5.63 Enthalpy of PA66

5.3.3 Polyethylene Terephthalate (PET)

Basic technical data

▶ Density: 1.350 to 1.370 g/cm³
▶ Melting point: 250 to 275 °C
▶ Glass transition temperature: 70 to 98 °C

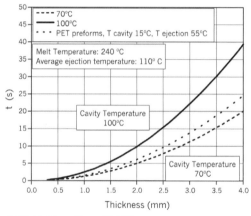

Figure 5.64 Mold cooling with PET

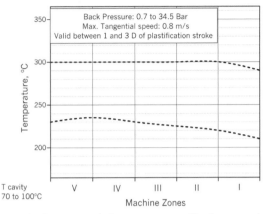

Figure 5.65 Recommended temperature profiles for processing PET

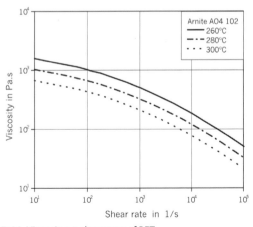

Figure 5.66 Viscosity vs shear rate of PET

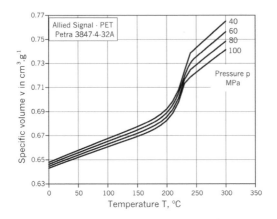

Figure 5.67 PVT diagram for PET

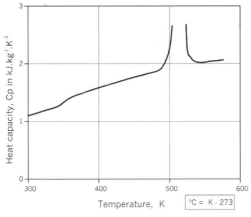

Figure 5.68 Heat capacity vs temperature of PET

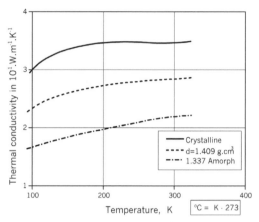

Figure 5.69 Conductivity vs temperature of PET

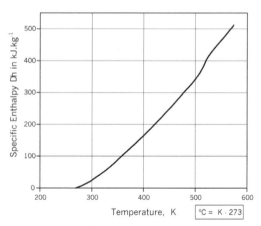

Figure 5.70 Enthalpy of PET

5.3.4 Polybutylene Terephthalate (PBT)

Basic technical data

▶ Density: 1.31 to 1.35 g/cm³
▶ Melting point: 190 to 250 °C
▶ Glass transition temperature: 45 to 60 °C

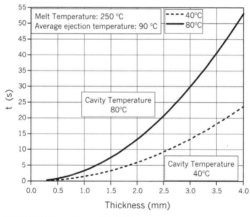

Figure 5.71 Mold cooling with PBT

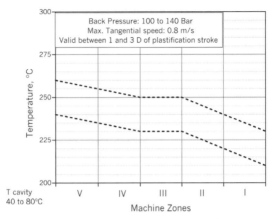

Figure 5.72 Recommended temperature profiles for processing PBT

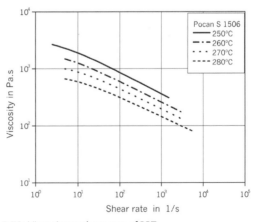

Figure 5.73 Viscosity vs shear rate of PBT

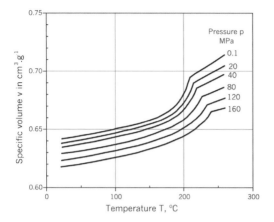

Figure 5.74 PVT diagram for PBT

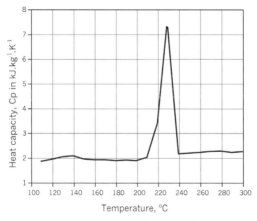

Figure 5.75 Heat capacity vs temperature of PBT

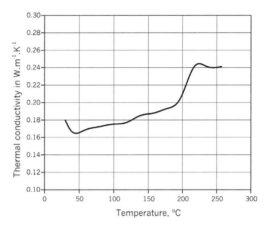

Figure 5.76 Conductivity vs temperature of PBT

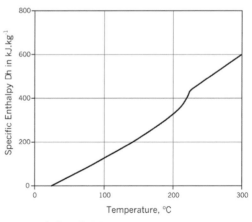

Figure 5.77 Enthalpy of PBT

5.3.5 Polycarbonate (PC)

Basic technical data

▶ Density: 1.200 g/cm^3
▶ Amorphous
▶ Glass transition temperature: 150 °C

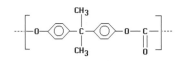

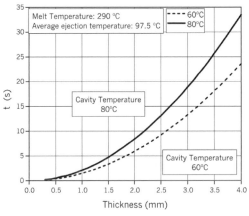

Figure 5.78 Mold cooling with PC

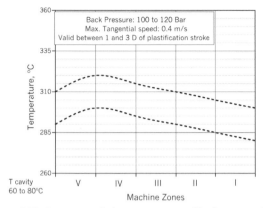

Figure 5.79 Recommended temperature profiles for processing PC

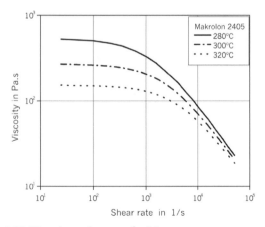

Figure 5.80 Viscosity vs shear rate for PC

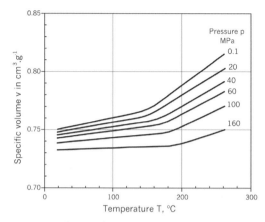

Figure 5.81 PVT diagram for PC

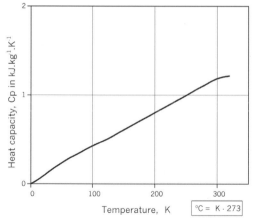

Figure 5.82 Heat capacity vs temperature of PC

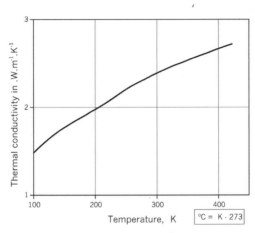

Figure 5.83 Conductivity vs temperature of PC

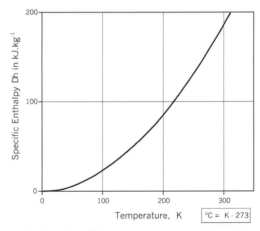

Figure 5.84 Enthalpy of PC

5.3.6 Polyphenylene Ether Modified (m-PPE)

Basic technical data

▸ Density: 1.060 g/cm^3
▸ Amorphous
▸ Glass transition temperature: 207 to 234 °C
▸ m-PPE = PPE+PS

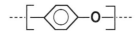

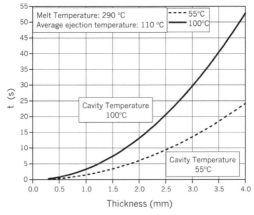

Figure 5.85 Mold cooling with PPE

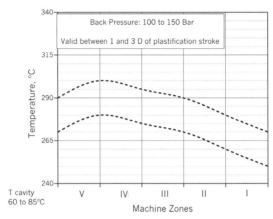

Figure 5.86 Recommended temperature profiles for processing PPE

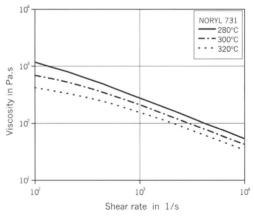

Figure 5.87 Viscosity vs shear rate of PPE

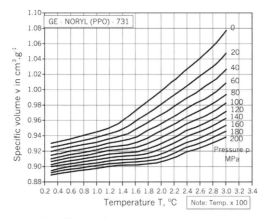

Figure 5.88 PVT diagram for PPE

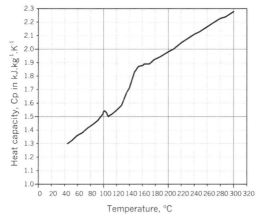

Figure 5.89 Heat capacity vs temperature of PPE

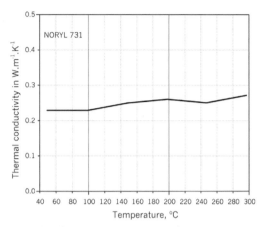

Figure 5.90 Conductivity vs temperature of PPE

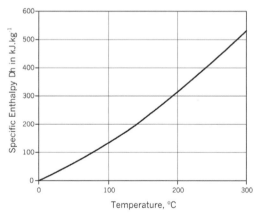

Figure 5.91 Enthalpy of PPE

5.3.7 Polyether Etherketone (PEEK)

Basic technical data

▶ Density: 1.265 to 1.320 g/cm^3
▶ Melting point: 256 °C
▶ Glass transition temperature: 143 °C
▶ Thermal conductivity: 0.249 W·m^{-1}·K^{-1}

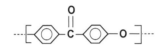

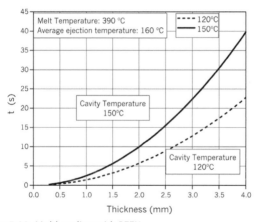

Figure 5.92 Mold cooling with PEEK

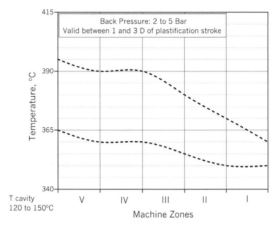

Figure 5.93 Recommended temperature profiles for processing PEEK

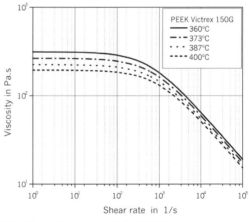

Figure 5.94 Viscosity vs shear rate for PEEK

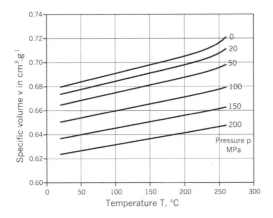

Figure 5.95 PVT diagram for PEEK

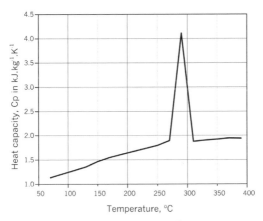

Figure 5.96 Heat capacity vs temperature of PEEK

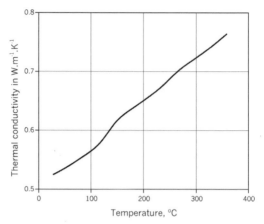

Figure 5.97 Conductivity vs temperature of PEEK

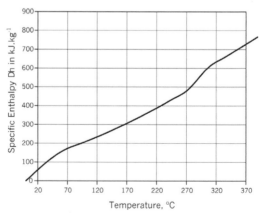

Figure 5.98 Enthalpy of PEEK

5.3.8 Polyarylsulfone (PSU)

Basic technical data

▶ Density: 1.240 to 1.450 g/cm³
▶ Amorphous
▶ Glass transition temperature: 190 °C
▶ Thermal conductivity: 0.26 W·m⁻¹·K⁻¹

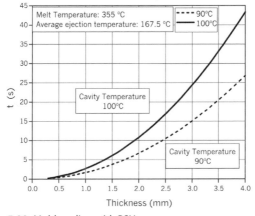

Figure 5.99 Mold cooling with PSU

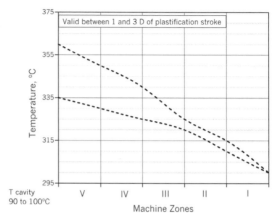

Figure 5.100 Recommended temperature profiles for processing PSU

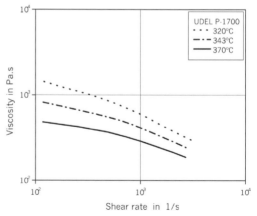

Figure 5.101 Viscosity vs shear rate of PSU

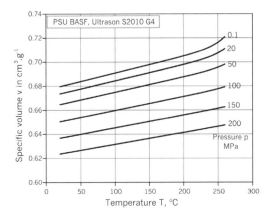

Figure 5.102 PVT diagram for PSU

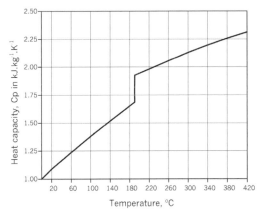

Figure 5.103 Heat capacity vs temperature of PSU

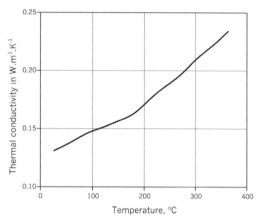

Figure 5.104 Conductivity vs temperature of PSU

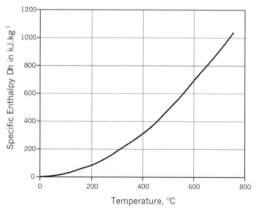

Figure 5.105 Enthalpy of PSU

5.4 Vinyls

5.4.1 Polyvinylchloride (PVC) Rigid

Basic technical data

▶ Density: 1.380 to 1.550 g/cm³
▶ Amorphous
▶ Glass transition temperature: 80 to 110 °C

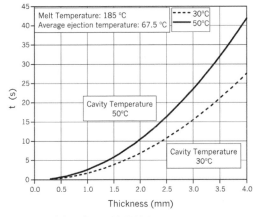

Figure 5.106 Mold cooling with PVC-R

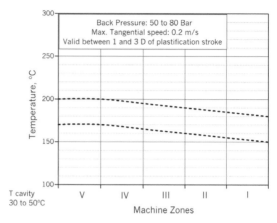

Figure 5.107 Recommended temperature profiles for processing PVC-R

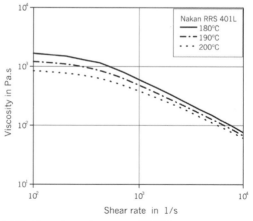

Figure 5.108 Viscosity vs shear rate of PVC-R

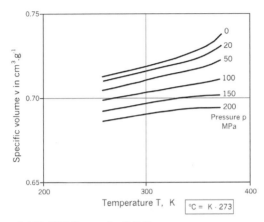

Figure 5.109 PVT diagram for PVC-R

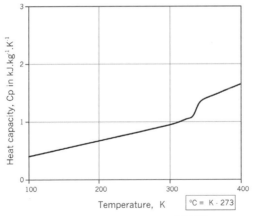

Figure 5.110 Heat capacity vs temperature of PVC-R

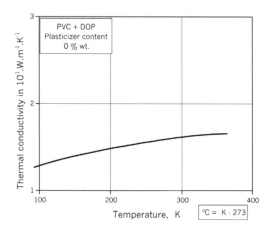

Figure 5.111 Conductivity vs temperature of PVC-R

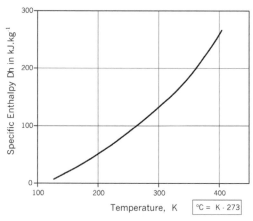

Figure 5.112 Enthalpy of PVC-R

5.4.2 Polyvinylchloride (PVC) Flexible

Basic technical data

▶ Density: 1.160 to 1.390 g/cm³
▶ Amorphous
▶ Glass transition temperature: −50 to 80 °C depending on the type and content of plasticizer

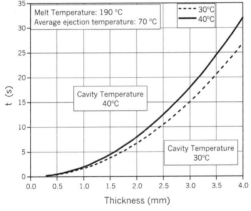

Figure 5.113 Mold cooling with PVC-F

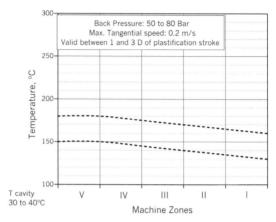

Figure 5.114 Recommended temperature profiles for processing PVC-F

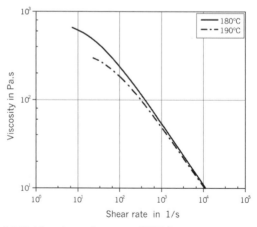

Figure 5.115 Viscosity vs shear rate of PVC-F

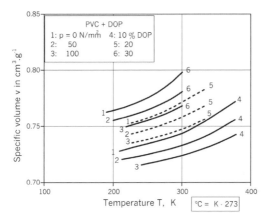

Figure 5.116 PVT diagram for PVC-F

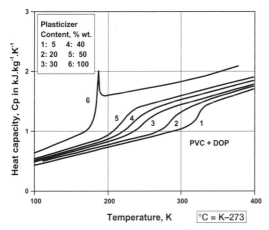

Figure 5.117 Heat capacity vs temperature of PVC-F

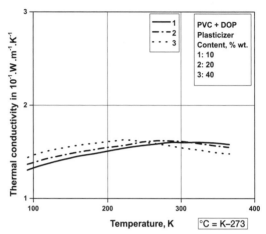

Figure 5.118 Conductivity vs temperature of PVC-F

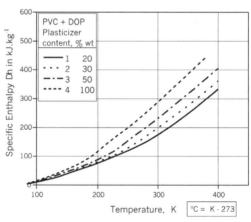

Figure 5.119 Enthalpy of PVC-F

5.5 Others

5.5.1 Polylactic Acid (PLA)

Basic technical data

▶ Density: 1.21 to 1.43 g/cm^3
▶ Melting point: 170 °C to 230 °C
▶ Glass transition temperature: 45 °C to 65 °C

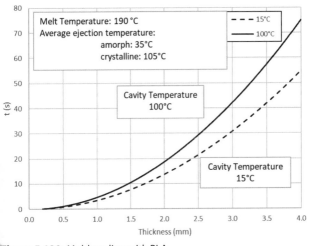

Figure 5.120 Mold cooling with PLA

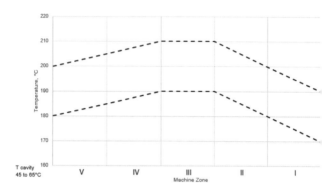

Figure 5.121 Recommended temperature profiles for processing PLA

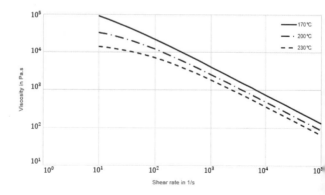

Figure 5.122 Viscosity vs shear rate of PLA

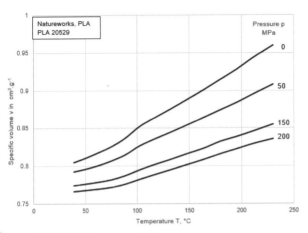

Figure 5.123 PVT diagram for PLA

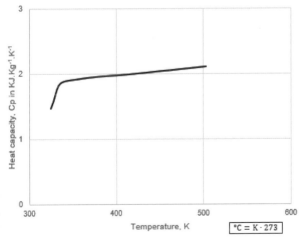

Figure 5.124 Heat capacity vs temperature of PLA

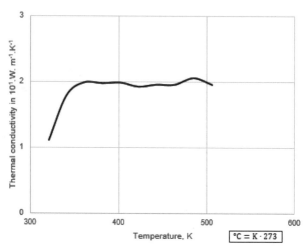

Figure 5.125 Conductivity vs temperature of PLA

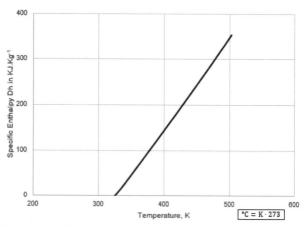

Figure 5.126 Enthalpy of PLA

5.5.2 Polymethylmethacrylate (PMMA)

Basic technical data

▶ Density: 1.120 to 1.200 g/cm^3
▶ Amorphous
▶ Glass transition temperature: 106 to 115 °C

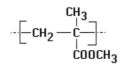

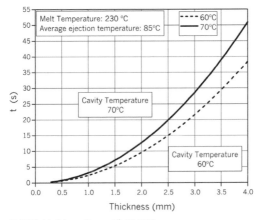

Figure 5.127 Mold cooling with PMMA

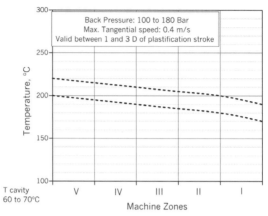

Figure 5.128 Recommended temperature profiles for processing PMMA

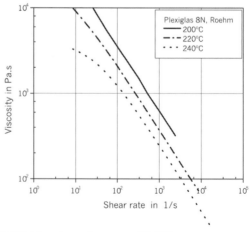

Figure 5.129 Viscosity vs shear rate of PMMA

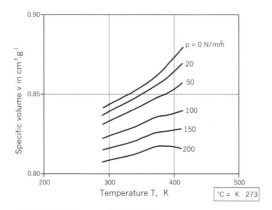

Figure 5.130 PVT diagram for PMMA

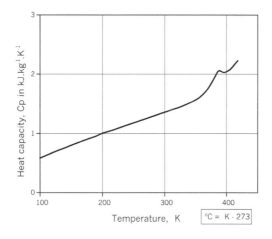

Figure 5.131 Heat capacity vs temperature of PMMA

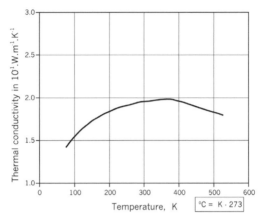

Figure 5.132 Conductivity vs temperature of PMMA

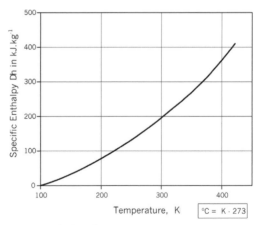

Figure 5.133 Enthalpy of PMMA

5.5.3 Polyacetal (POM)

Basic technical data

▶ Density: 1.410 to 1.420 g/cm^3
▶ Melting point: 164 to 168 °C (copolymer); 175 to 181 °C (homopolymer)
▶ Glass transition temperature: −85 to −73 °C

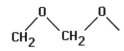

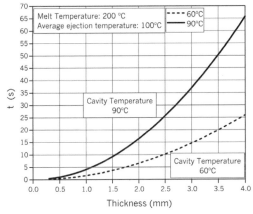

Figure 5.134 Mold cooling with POM

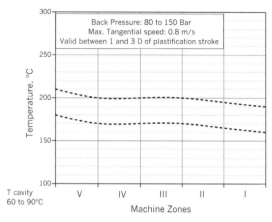

Figure 5.135 Recommended temperature profiles for processing POM

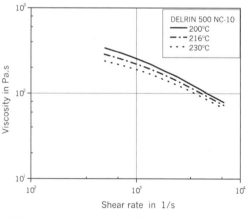

Figure 5.136 Viscosity vs shear rate of POM

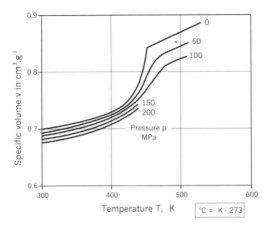

Figure 5.137 PVT diagram for POM

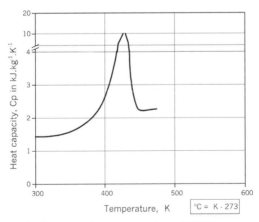

Figure 5.138 Heat capacity vs temperature of POM

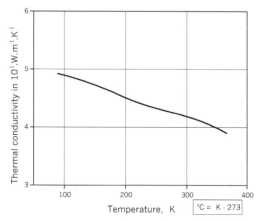

Figure 5.139 Conductivity vs temperature of POM

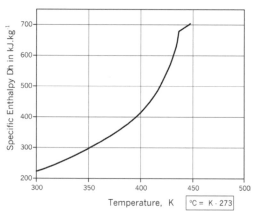

Figure 5.140 Enthalpy of POM

6 Appendix

Table 6.1 Processing Window and Average Linear Shrinkage for Various Thermoplastics Polymers

Polymer	Max. shear rate, 1/s	Melt temperature	Average shrinkage
ABS	45,000	220 to 260 °C	0.50%
PS/HIPS	55,000	180 to 280 °C	0.45%
SAN	45,000	220 to 260 °C	0.50%
LDPE	65,000	180 to 200 °C	2.60%
HDPE	50,000	190 to 280 °C	2.40%
PP	85,000	230 to 260 °C	2.00%
PA66	65,000	280 to 300 °C	1.40%
POM	65,000	210 to 220 °C	1.90%
PBT	55,000	250 to 260 °C	1.90%
PET	9,000	280 to 300 °C	1.20%
PC	35,000	290 to 320 °C	0.50%
PMMA	21,000	220 to 260 °C	0.50%
U-PVC	5,600	180 to 215 °C	0.4%

Table 6.2 Drying Conditions for Hygroscopic Materials

Material	Drying tempe-rature (°C)	Drying time (hours)	Max. water content (%)
PET	120 to 160	4 to 6	< 0.02
PETG	65	4 to 6	< 0.02
PLA	65 to 90	5 to 8	< 0.02
CA	65 to 75	2 to 4	< 0.02
PA6	75 to 85	2 to 20	0.1 to 0.2
PA11	65 to 80	3 to 4	0.1 to 0.2
PA12	70 to 80	3 to 4	0.1 to 0.2
PA66	60 to 80	2 to 20	0.1 to 0.2
SAN	70 to 90	1 to 4	0.05 to 0.1
ABS	80 to 100	2 to 6	0.05 to 0.2
ASA	80 to 85	2 to 4	0.2
PBT	110 to 140	2 to 8	0.02 to 0.05
PC	120	2 to 12	0.01 to 0.02
PPE+HIPS	80 to 120	2 to 4	0.05 to 0.1
PMMA	70 to 100	8	0.05 to 0.1
PEEK	150 to 160	2 to 3	0.1
PSU	135 to 163	2 to 3.5	0.05
POM	80 to 120	4	0.05 to 0.1

Table 6.3 Cavity and Ejection Temperature for Selected Materials

Material	Cavity temperature (°C)	Ejection temperature (°C)
PS	10 to 30	35 to 45
HIPS	10 to 30	35 to 50
SAN	40 to 50	50 to 70
ABS	40 to 55	55 to 75
PVC (rigid)	30 to 50	65 to 70
PVC (flexible)	30 to 40	35 to 45
PMMA	60 to 70	70 to 100
PPE+HIPS	55 to 100	75 to 140
PC	60 to 80	85 to 110
SU	120 to 150	160 to 175
LDPE	10 to 30	40 to 50
HDPE	10 to 40	50 to 60
PP	30 to 45	55 to 65
PA6	60 to 80	70 to 90
PA66	60 to 100	90 to 120
POM	60 to 90	85 to 115
PET	70 to 100 (10 to 30 preforms)	95 to 125 (preforms 55)
PBT	40 to 80	80 to 100
PEEK	120 to 150	145 to 160

7 Definitions

Cavity (of a mold): the space within a mold to be filled to form the molded product.

Cavity temperature: the temperature of the surface in contact with the polymer melt. This temperature is different from the cooling liquid temperature.

Channel depth: one half the difference between the external diameter and the root diameter of the screw.

Cooling time: the time interval from the start of packing time until the mold starts to open.

Degree of cooling: defined by the following ratio, where T_M is the melt temperature, T_W stands for the mold cavity wall temperature, and T_E for the average temperature across the thickness of the part at demolding:

$$Degree\ of\ cooling = \frac{T_M - T_W}{T_E - T_W}$$

Density: the weight per unit volume of a material.

DOP: dioctyl phthalate.

Effective thermal diffusivity: the value of thermal diffusivity that is independent of the material temperature (slightly dependent on the mold), which produces a certain degree of cooling. Symbol: α_{eff}.

Ejection temperature: average temperature at which the molded product is removed from the cavity.

Enthalpy: the difference in heat content of a substance between two temperatures, T_1 and T_0, where T_0, the reference temperature, is usually 0 or 20 °C.

Extrusion: a process whereby heated or unheated plastic is forced through a shaping orifice and becomes one continuously formed piece.

Glass transition: the reversible change in an amorphous polymer or in amorphous regions of a partially crystalline polymer from (or to) a viscous or rubbery condition to (or from) a hard and relatively brittle one.

Glass transition temperature: the approximate midpoint of the temperature range over which the glass transition takes place. *NOTE:* the glass transition temperature (T_g) varies significantly, depending upon the specific property tested, the test method, and conditions selected to measure it.

Heat capacity: the amount of heat necessary to increase the temperature of 1 g of substance by 1 K at a constant pressure.

Injection molding: the process of molding a material by injection under pressure from a heated cylinder through a sprue (runner, gate) into the cavity of a closed mold.

Melt temperature: temperature of the molten plastic.

Melting point: the temperature, measured under specified test conditions, at which crystallinity disappears in a semi-crystalline polymer on heating.

Mold: an assembly of parts enclosing the space (cavity) from which the molding takes its form.

Plasticize: to render a polymeric material softer, more flexible, and/or more workable by the addition of a plasticizer or by chemical modification of the polymer.

Plasticizer: a substance of a low or negligible volatility incorporated in a plastic to lower its softening range and to increase its workability, flexibility, or extensibility.

Processing temperature: the temperature that allows for a low viscosity for adequate processing.

Representative shear rate: the value of shear rate for a non-Newtonian fluid at a certain point of the channel, which is coincident with the value for a Newtonian fluid for the same volumetric flow.

Runner: (1) the secondary feed channel in an injection or transfer mold that runs from the inner end of the sprue to the cavity gate; (2) the molding material in this secondary feed channel.

Shear rate $\dot{\gamma}$ (s^{-1}): the time rate of change of shear strain:

$$\dot{\gamma} = \frac{d\gamma}{dt}$$

Shear strain γ (dimensionless): the tangent of the angular change, due to force, between two lines originally perpendicular to each other through a point in a body.

Specific volume: the volume per unit weight of a material.

Temperature gradient: change of temperature per change of length.

Thermal conductivity (of a homogeneous material not affected by thickness): the rate of heat flow under steady conditions through a unit area, per unit temperature gradient in the direction perpendicular to the area. IUPAP symbol: λ.

Thermal diffusivity: the ratio of the thermal conductivity of a substance to the product of its density and specific heat. IUPAP symbol: α. This property is defined only for an unsteady state process and indicates the speed at which thermal energy is propagated in a body subjected to a temperature change.

Thermoplastic: material capable of being repeatedly softened by heating and hardened by cooling through a temperature range characteristic of the plastic.

Viscosity η (Pa·s): resistance offered by a material under a deformation.

8 Further Reading

CAMPUS Database Software version 4.0 and 4.1

Domininghaus, H., *Plastics for Engineers: Materials, Properties, Applications* (1993), Hanser Publishers, Munich

Osswald, T.A. and Menges, G., *Materials Science of Polymers for Engineers*, 3rd ed. (2012), Hanser Publishers, Munich

Osswald, T.A., Gramann, P.J., and Turng, L.S., *Injection Molding Handbook*, 2nd ed. (2007), Hanser Publishers, Munich

Pötsch, G. and Michaeli, W., *Injection Molding: an Introduction*, 2nd ed. (2007), Hanser Publishers, Munich

Stevenson, J.F., Polymer Engineering & Science. *Society of Plastics Engineers* 18, 577 (1978)

Van Krevelen, D.W. and Te Nijenhuis, K., *Properties of Polymers: Their Correlation with Chemical Structure; Their Numerical Estimation and Prediction from Additive Group Contributions*, 4th ed. (2009), Elsevier, ISBN 978-0-08-054819-7

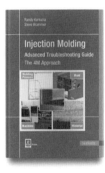